YOUR KNOWLEDGE HAS VALUE

- We will publish your bachelor's and master's thesis, essays and papers

- Your own eBook and book - sold worldwide in all relevant shops

- Earn money with each sale

Upload your text at www.GRIN.com and publish for free

Pratibha Chaturvedi, Abhay Chowdhary

Plant derived vaccine

Edible vaccine

GRIN Verlag

Bibliografische Information der Deutschen Nationalbibliothek:

Die Deutsche Bibliothek verzeichnet diese Publikation in der Deutschen National-
bibliografie; detaillierte bibliografische Daten sind im Internet über http://dnb.d-
nb.de/ abrufbar.

Imprint:

Copyright © 2014 GRIN Verlag GmbH
Druck und Bindung: Books on Demand GmbH, Norderstedt Germany
ISBN: 978-3-656-60664-2

This book at GRIN:

http://www.grin.com/en/e-book/269489/plant-derived-vaccine

PLANT DERIVED VACCINE: A REVIEW

BY-

DR PRATIBHA CHATURVEDI (RESEARCH ASSOCIATE)

DR (PROF) ABHAY CHOWDHARY(DIRECTOR)

HAFFKINE INSTITUTE FOR TRAINING,RESEARCH AND TESTING

PAREL ,MUMBAI 400012

2014

ABSTRACT

Biotechnologists in recent years have come up with a new concept. This new concept is about edible vaccine. Edible vaccines are composed of antigenic proteins and do not contain pathogenic genes (because obviously they use attenuated strains). Thus, they have no way of establishing infection and safety is assured. Edible vaccines hold great promise as a cost-effective, easy-to-administer, easy-to-store, fail-safe and socio culturally readily acceptable vaccine delivery system, especially for the poor developing countries. It involves introduction of selected desired genes into plants and then inducing these altered plants to manufacture the encoded proteins. Resistance to genetically modified foods may affect the future of edible vaccines. This Review article include many aspects related to plant derived vaccine like Molecular farming, Process of development of plant derived vaccines. Mechanism of action .recent research regarding development of edible vaccines against cholera, malaria, Hepatitis B, rabies etc. The review article gives a overall picture of Plant derived vaccine.

Key words: Plant, Vaccines, Protein, Antigen, Diseases, Tissue culture

Introduction

The plant-based vaccine production method works by isolating a specific antigen protein, one that triggers a human immune response from the targeted virus. A gene from the protein is transferred to bacteria, which are then used to "infect" plant cells. The plants then start producing the exact protein that will be used for vaccinations. The flexibility of the plant expressed vaccine system, combined with its low cost and ability to massively scale, may provide vaccine protection not only to citizens of the United States, but to many parts of the world that cannot currently afford vaccines. Subunit vaccines that consist of one or more antigenic epitopes or proteins are often preferred to traditional vaccines made of killed or attenuated organisms. Mammalian, yeast and insect cell cultures are used to produce subunit vaccines because of their ability to process recombinant proteins in a manner similar to that of the native organism. However, expensive media and the purification steps needed for recovering recombinant proteins expressed in these organisms increase the cost of producing these subunit vaccines. In addition, most subunit vaccines produced in these systems are heat sensitive and require parenteral delivery. This restricts use of subunit vaccines in the poorly funded health systems of developing countries.

A promising alternative is to transform plants with a gene(s) encoding an immunogenic protein capable of preventing infection by a pathogenic agent. The production of vaccines in transgenic plants overcomes the risk of contamination with mammalian pathogens and can enable oral delivery. These characteristics simplify vaccine delivery and decrease the cost of an immunization program.

History of plant derived vaccine

By the late 1990s an international campaign to immunize all the world's children against six devastating diseases was reportedly reaching 80 percent of infants (up from about 5 percent in the mid-1970s) and was reducing the annual death toll from those infections by roughly three million. Vaccines trigger and prepare our body's defense mechanisms so that the system is able to fight and eliminate the pathogens when encountered due to natural infection. Recent progress in vaccine technology has improved public health to a remarkable extent. Each year, millions of children in underdeveloped countries have no access to immunization. The traditional vaccines are expensive and require special conditions for storage, distribution and & dispensing. Moreover, the supplies of traditional vaccines are limited and they are in short supply. The first report of the production of edible vaccine (a surface protein from *Streptococcus*) in tobacco, at 0.02% of total leaf protein level, appeared in 1990 in the form of a patent application published under the International Patent Cooperation Treaty. Fifteen years later, the first technical proteins produced in transgenic plants are on the market, and proof of concept has been established for the production of many therapeutic proteins, including antibodies, blood products, cytokines, growth factors, hormones, recombinant enzymes and human and veterinary vaccines. Subsequently, a number of attempts were made to express various antigens in plants[10-12].Arntzen and John Clements, of Tulane University Medical School, launched an immunological battle against gut-invading bugs in 1991 using "bio-pharmed" tobacco to target a form of *Escherichia coli* (*E-coli*), a diarrhea-causing bacterium that kills approximately three million infants each year. In the past five years experiments conducted by Arntzen (who moved to the Boyce Thompson Institute for Plant Research at Cornell University in 1995) and his collaborators have demonstrated that tomato or potato plants can synthesize antigens from the Norwalk virus, Enterotoxigenic, E. coli, V.cholerae and the Hepatitis B .

Molecular farming

Plant biologists had developed a plan of introducing selected genes (the blueprints for proteins) into plants and inducing the manipulated, or transgenic, plants to manufacture the encoded proteins. For making edible vaccines against the different pathogens, it is necessary to find out pathogen associated antigenic epitopes or surface antigens. The antigenic epitopes are proteins or peptides that are encoded by genomic sequences. The basic methodology includes identification, selection and isolation of desirable genes from the pathogen that encodes the surface antigen proteins. The isolated gene can be then cloned in a suitable vector for gene transfer. The selected vector should possess all the unique characteristics of an ideal vector. The molecular markers present in vectors can be used for screening transformed host cells from untransformed. After integration of desirable gene in host genome, the cells can be checked for cloned gene expressions using eliza that ultimately uses antigen specific monoclonal antibodies. The transformed cells with positive cloned gene expressions allow them for propagation using plant tissue culture (Subbirhussain, 2011).

As molecular farming has come of age, there have been technological developments on many levels, including transformation methods, control of gene expression, protein targeting and accumulation, the use of different crops as production platforms (Twyman *et al*, 2003,2005), and modifications to alter the structural and functional properties of the product. One of the most important driving factors has been yield improvement, as product yield has a significant impact on economic feasibility. Strategies to improve the recombinant protein yield in plants include the development of novel promoters, the improvement of protein stability and accumulation through the use of signals that target the protein to intracellular compartments, and the improvement of downstream processing technologies. [13]

Considering the cost of protein purification as comparable savings in the upstream components make the production cost of a commercially important protein in plants substantially less than other systems. Genetically modified plants can be grown in large area. [14,15].The cost of goods sold(COGS) for bulk production of recombinant protein in plants has been estimated to be 1/10th to 1/50th of bacteria lfermentation[16].Therefore it is economically sound to use transgenic plants for antigen production. A variety of gene expression and protein localization systems now available for plants allows stable accumulation of the recombinant proteins in target plant tissue (Fig.1.).A plant ideal for oral vaccine production should have the following features:

(i)   A men ability to transformation,

(ii)   Expression in edible tissue that can be consumed un cooked since vaccine antigens are heat sensitive

(iii)   Targeted tissue to be rich in protein because vaccine protein will only be a small percentage of the total protein.

(iv)   Targeted tissue should not produce toxic molecules and would allow correct folding of the antigen protein and desired post translational modifications. Expression of antigen in edible tissue offers a convenient and inexpensive source to deliver a vaccine.

Expression of commercially important proteins in leaf tissue is not a good strategy[17,18,19] on account of the following reasons:

(i)Over all protein content in leaf tissue is low.

(ii) Leaves have high protease activity.

(iii)The presence of pigments and phenolic makes purification of recombinant protein from leaves

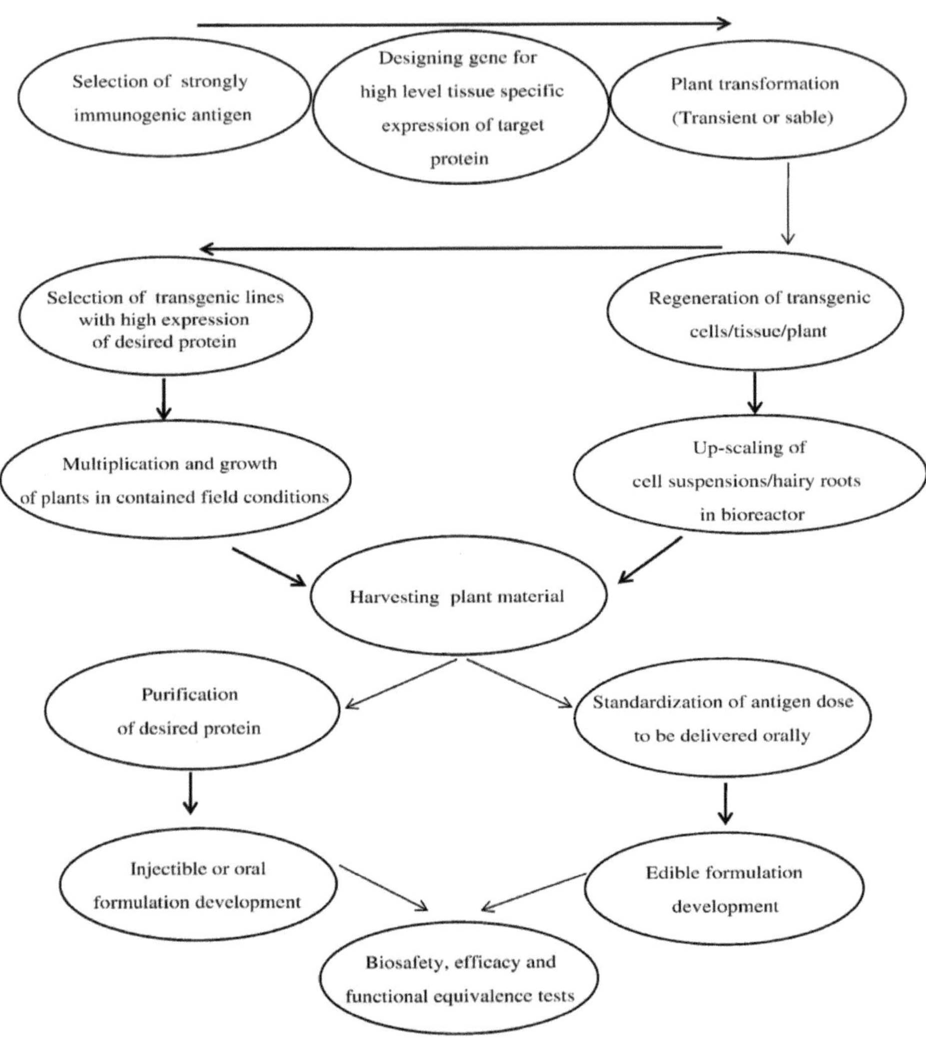

Figure 1.Curtesy :Tiwari et al.,2009.**Steps in the production of plant-derived vaccine**

Manufacturing practices becomes more difficult due to the need for handling large volume and biomass. The expression levels of plant-derived bio-pharmaceuticals need to be increased before commercial production would accomplished[20,21] on economically competitive basis. The expression levels of their combinant proteins in the transgenic plants are also influenced by environmental factors. High expression levels could be best achieved in cell suspension, hairy root cultures (in vitro) and seeds(in vivo).Seed tissue represents potentially a very promising target for producing pharmaceutically important proteins for extraction at commercial level. There combinant seeds also offer the possibility of direct use as an edible vaccine. Single chain antibodies expressed in seeds of rice and wheat showed high biological activities and remained stable for several years [22].Thus,the proteins expressed in seeds are highly stable. Long term storage and easy transportability of seeds is possible due to very low moisture content of mature seeds[23] Other tissues like hairy root and cell suspension cultures could be useful target issues to express recombinant proteins[24-26],though the establishment and running costs of such in vitro systems are higher.

One of the most important aspects in molecular bio-farming is the selection of promoter to achieve high level expression of the antigen coding gene. The choice of promoters affects trans gene transcription, resulting in change snot only in concentration, but also in the stage, tissue and cell specificity of its expression. Cauliflower mosaic virus 35S(CaMV35S)promoter has widely been used because of its strong and constitutive expression. High level protein expression is essential to develop economically competitive plant-based process for cultivation of the transgenic variety within confines fields with controlled environmental and contained biosafety conditions highly expressing and yet tightly controlled promoters are desirable for bio-farming proteins from plants[27].Expression of a protein with CaMV35S promoter does not permit regulated gene expression. Though CaMV35S promoter expresses genes at a relatively high level in leaves and roots, low level of total protein .Many therapeutic proteins can be expressed in stable or transient state in whole plants, plant tissues or cell suspension cultures. There is growing acceptance of transgenic crops in both developed and developing countries. The future of edible vaccines depends on acceptability for genetically modified foods. Successful implementation of edible vaccines depend on how well we overcome various technical complications, regulatory issues and non-scientific challenges. .The World Health Organization (WHO) has called for new strategies to deliver vaccines. WHO estimates that 10 million children die in developing countries each year from infectious diseases that could be prevented with vaccines.[28]

Approach for Production of Edible Vaccines

Creating edible vaccines involves introduction of selected desired genes into plants or animal and then inducing these altered plants or animals to manufacture the encoded proteins [29,30]. This process is known as "transformation," and the altered plants or animals are called "transgenic plants" or "transgenic animals". Edible vaccines are similar to conventional subunit vaccines as they are composed of antigenic proteins andare devoid of pathogenic genes, Hence, they cannot cause infection and can be safely used in patients with weak immune system. Immunization is done by feeding animals or humans with food derived from edible parts of transgenic plants or animals in which an orally active antigen of the target pathogen is expressed and accumulated. Immunoglobulin molecules have been successfully synthesized in tobacco plants using the same technology

The desired gene which codes for antigen is isolated from the microbes and could be handled in two different ways:

1.Appropiate plant virus is genetically engineered to express thedesired peptidesproteins. The recombinant viru s is then inoculated into the plant.The resultant plant edible vaccines are utilized for immunological purpose

2.In another approach, the gene of interest is integrated with plant vector by transformation Avariety of techniques have been used to introduce transgene into plant cell.

These could be grouped into following categories:

A: Agro bacterium mediated gene transfer:

The appropriate gene construct is inserted into theTregion of a disarmed Ti plasmid of Agrobacterium. The rec ombinant DNA is placed into Agrobacterium; a plant pathogen which is cocultured with the plant cells or tissu es to be transformed.[31] The drawback of this method is that it gives low yield and the process is slow. Thismet hod works especially for dicotelydenous plants like potato, tomato and tobacco. Studies have also proved that the genes are expressed by this method in experimental animals and plants.[18-20]

B: Biolistic method:

The gene containing DNA coated metal (e.g. gold, tungsten) particles are fired at the plant cells using gene gun.21 Those plant cells that take up the DNA are then allowed to grow in new plants, and are cloned to produce large number of genetically identical crop. This method is quite attractive because DNA can be delivered into cells of plant which makes gene transfer independent of regeneration ability of the species. But the chief limitation is the need for costly device particle gun.

C: Electroporation:

Here there is introduction of DNA into cells by exposing them for brief period to high voltage electrical pulse, which is thought to induce transient pores in the plasma lemma.[32] The cell wall presents an effective barrier to DNA therefore, it has to be weakened by mild enzymatic treatment so as to allow the entry of DNA in to cell cytoplasm.

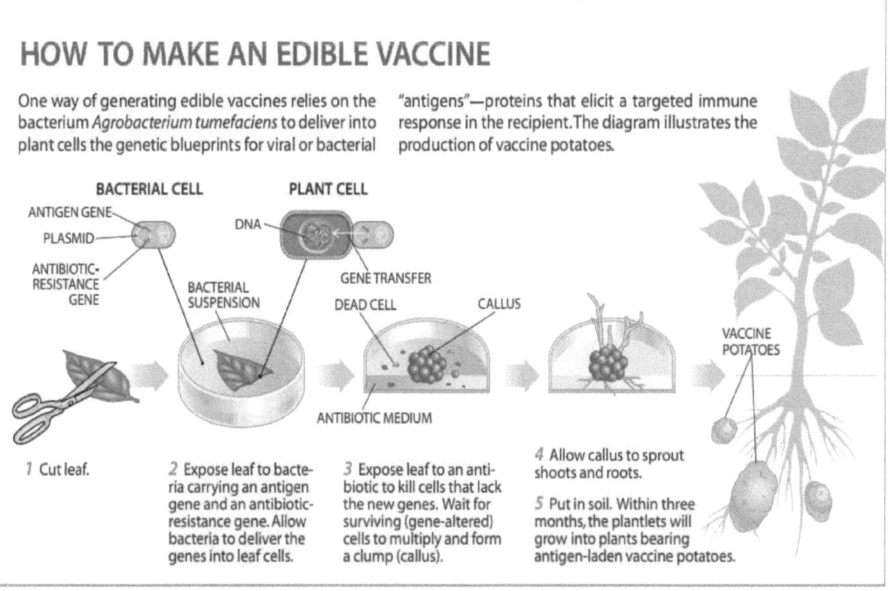

Courtesy William H. R., 2000

Figure 2. Depicts the process involved in plant derived vaccine production

Mechanism of action

1.The first line of the defense mechanism is mucosal immunity. Vast majority of human pathogens initiates infections at mucosal surfaces, by making gastrointestinal, urogenital and respiratory tracts as major routes of entry into the body.[32]

2.The most effective route of mucosal immunization is oral route.The oral vaccine can generate mucosal immunity, antibody mediated immune response and cell mediated immune response. The antigen containing plant vaccine when introduced orally does not getdegradedby gastric enzymes because of the presence of tough outer wall of plant cell. Release of antigen and its breakdown from plant vaccine occurs near Payer's patches of intestine.[34] The released antigen is taken up by the Mcells and is presented to B cells with the help of antigen presenting cells (APC). The activated Bcells get differentiated into plasma cells and secrete Ig A class of antibody and elicit mucosal immunity and humoral immunity. Another important component of mucosal

Immunity is T
cell mediated immune response where the Tcells specifically recognize pathogens and directly kill the invader themselves. It also helps indirectly to the antibodies to clear infection. Tcells produced in
the mucous are capable of travelling the mucosal tissues through special 'homing' receptors on their membranes. This means that if an immune response is generated in gastrointestinal lining, Tcells produced there can travel to other mucosal sites, (e.g. the lungs, nasal cavity) providing protection over a large surface area. [35]

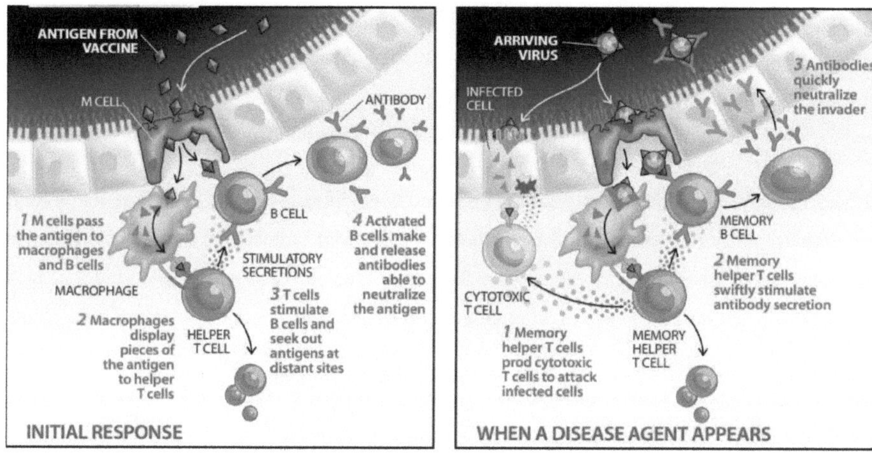

Figure3. showing the mechanism of action of edible vaccine Courtesy William H. R. ,2000

Advantages of plant derived vaccine

1.Plant derived vaccines are effective as delivery vehicle for immunization because adjuvants which enhance the immune response are not necessary.

2. They can elicit mucosal immunity which is not observed in traditional vaccines .

3.Edible vaccines are also cost effective in storage, preparation, production and transportation.

4.Vaccines produced by biotechnological method are stable at room temperature, unlike traditional vaccine w hich need cold chain storage which increases the yearly cost to preserve vaccines.[36] Moreover, the seeds of transgenic plants could be dried as there is less moisture content in seeds and the plants with oil or th eir aqueos extracts possess more storage opportunities.[37] For production of edible vaccines, costly equipments and machines are not necessary as they could be easily made on soil rich motherland and the cost for growing plants is also low compared to cell culture grown in fomenters.

5.As edible vaccines are produced from plants, they are easily available. Manufacturing cost is low as there is no need of special premises for manufacturing them and facilities like sterilization as required in traditional vaccines.

6.Edible vaccines are easily acceptable as they do not require administration by injection unlike traditi onal vaccines. Thus there is also reduced need of medical personnel and the contamination risk is low as sterile conditions are not necessary.

The risk of spreading environmental contamination by spreading second hand diseases is minimized.

7.The Plant based vaccines are cheaper especially for the poor and developing countries. . Because many food plants can be regenerated readily, the crops could potentially be produced indefinitely without the growers having to purchase more seeds or plants year after year. The technique for production of edible vaccines involves introduction of selected desired genes of immunogenic proteins from various pathogens into plants and animals and then inducing these altered organisms to manufacture the encoded proteins whereas existing vaccines are expensive and require a semi-skilled person to give the injection with needles that are hard to come by in developing countries. Reused needles can transmit viruses such as Hepatitis B and C and HIV.

8. They are easy to administer and show better compliance, especially in children and do not require trained medical personnel. A large number of efficient, large scale production processes are being developed for them. They are administered without purification, reducing production cost and can be easily grown locally without requirement of capital intensive pharmaceutical manufacturing facilities

9. Edible vaccines show appreciable level of genetic and thermal stability eliminating the need of maintenance of refrigerated conditions during transportation and storage.

10.Plant derived vaccines could be the source for new vaccines combining numerous antigens. These multicom ponent vaccines are called second generation vaccines as they allow for severalantigens to approach M cells si multaneously.

11.An edible vaccine improves the safety of individual as compared to traditional vaccine since there is no pos sibility of proteins to reform into infectious organism. Edible vaccines are subunit preparation and do not invol ve attenuated pathogens.

12Edible vaccine is easy for mass production system by breeding as compared to an animal system.

Limitations of plant derived vaccines

Plantderived vaccine is easy to access but still it leaves some challenges. The major problems associated with e dible vaccine are as follows:

1. Possibility of development of immune tolerance to the vaccine protein or peptide.

2.Consistency of dosage form differs from plant to plant and generation to generation, protein content, patient's age, weight, ripeness of the fruit and quantity of the food eaten in absenceof availability of methods for standardization of plant material/product. Low doses of the edible vaccine if consumed produces less number of antibodies and high doses causes immune tolerance.

3. Stability of vaccine differs from plant to plant.

4.Some food cannot be eaten raw (e.g. potato) and needs cooking which will denature or weaken the protein present in it.

5.Variable conditions for edible vaccine are also a major problem. Potatoes containing vaccine to be stored at 4 c and could be stored for longer time while a tomato does not last long. Thus these vaccines need to be properly stored to avoid infection through microbial spoilage.

6.Another concern regarding edible vaccine is need of proper distinguishing characters to identify between 'vaccine fruit' and 'normal fruit' to avoid misadministration of vaccine which could lead to tolerance.

7. Glycosylation pattern of plants and humans is different which could affect the functions of vaccines.

Research regarding edible vaccine

Cholera

An attempt was made towards the production of plant derived vaccine by expressing heat-labile enterotoxin (LT-B) in tobacco and potato[37].The enterotoxin (LT) from *E. coli* is a multimeric protein, quite similar to cholera toxin (CT) structurally, functionally and antigenically. LT has one A subunit (27 kDa) and a pentamer of B subunits (11.6 kDa). Binding of the non-toxic LT-B pentamer to GM_1 gangliosides, present on epithelial cell surfaces, allows entry of the toxin LT-A subunits into the cells. LT-B and CT-B are both potent oral immunogens. An oral vaccine composed of the cholera toxin-B subunit (CT-B) with killed *V. cholerae* cells has been reported to give significant level of protection against cholera[8]. But the cost of production of CT-B by conventional methods is too high to allow distribution of this vaccine. The recombinant LT-B (rLT-B) produced in tobacco and potato showed partial pentamerization after the engineering of subunit gene in a way that allowed retention of the protein in microsomal vesicles. On testing immunogenicity of rLT-B by feeding potato tubers to mice, both humoral and mucosal immune responses were reported to be stimulated. This vaccine has gone through pre-clinical trials in humans. The antigenic protein retained its immune genecity after purification from the transgenic potato expressing it[27]. Fourteen healthy individuals, who ate 50–100 g raw potatoes, were screened for gut-derived antibody secreting cells (ASC), which were detectable 7–10 days after immunization. Presence of both anti-LT IgA-secreting cells and anti-LT IgG-secreting cells was detected in the peripheral blood.Cholera toxin, which is very similar to *E. coli* LT, has also been expressed in plants. Plants expressing CT-B showed the presence of a protein that migrated to the same position in denaturing gel as the CT-B derived from *V. cholerae*, and was recognized by mouse anti-CT-B antibody. Cholera toxin-B subunit, when expressed in potato, was processed in a natural way: the pentameric form (the naturally occurring form) being the abundant form. Antigenically it was found to be similar to the bacterial protein. Even after boiling transgenic potato tubers till they became soft, approximately 50% of the CT-B was present in the pentameric GM_1 ganglioside-binding form. LTB is a potent mucosal immunogen and is commonly used as an adjuvant to stimulate antibody response, when co-expressed with other antigens. LTB and its genetic fusions with other proteins have been successfully expressed in tobacco, potato, maize, tomato, *Arabidopsis thaliana*, soybean and carrot[37,38,39].

Rabies

A rabies virus coat glycoprotein gene has been expressed in tomato plants. The protein that was expressed had molecular mass of 62 kDa compared to 66 kDa observed from virus grown in BHK cells. Since the orally administered protein elicited protective immunity in animals, it was expected that continued efforts would lead to development of an edible oral vaccine against rabies which could be used as a preventive strategy. While the results with antigenic properties of the components produced in plants are encouraging, their value as a vaccine can be improved by providing other adjuvants which either enhance the immunogenic potential or reduce the degradation of the active ingredient by the gut microflora.(Table)

Hepatites B

Plants were considered as alternative sources of vaccines, to be mainly orally administered. Despite20-year attempts ,no real anti-HBV plant-based vaccine has been developed immunization trials, based on ingestion of raw plant tissue and conjugated with injection or exclusively oral administration of lyophilized tissue, were either impractical or insufficient due to oral tolerance acquisition.(TomaszPniewski,2013)The Hepatitis

B surface antigen (HBsAG) has been reported to accumulate to 0.01% of soluble protein level in transgenic tobacco[3]. The antigens, delivered in a macromolecular form, are known to survive the gut atmosphere and perform better. A crude extract from plants was used for parenteral immunization in mice. The immune response included all IgG subclasses as well as IgM against hepatitis B[40]. Carrillo *et al.*[41]expressed structural protein, VP1, of foot-and-mouth disease virus in *Arabidopsis*.

One of the alternative strategies of producing a plant-based vaccine is to infect the plants with recombinant viruses carrying the desired antigen that is fused to viral coat protein. The infected plants have been reported to produce the desired fusion protein in large amounts in a short time. The technique involved either placing the gene downstream a subgenomic promoter, or fusing the gene with capsid protein that coats the virus . The latter strategy is perhaps the strategy of choice since fusion proteins in particulate form are highly immunogenic. It should, however, be kept in view that recombinant viruses need to be highly purified for parenteral administration or partially purified for oral administration.

.

Malaria

Malaria remains one of the most significant causes of human morbidity and mortality worldwide, with 300 to 500 million new cases of infection annually resulting in 1.5 to 2.7 million death. The world malaria situation has become significantly worse in recent years as the main forms of malaria control,. Three antigens are currently being investigated for the development of a plant vaccine merozoite surface protein(MSP) 4 and MSP 5 from Plasmodium falciparum and MSP4/5 from *P.yoelli*. for those study ,however protein were expressed in E.coli‌Plant expression systems have been poorly explored for development of vaccines against human parasite pathogens. In fact, only few antigens from Plasmodium ssp. and Toxoplasma gondii have been expressed in plants [42,43].

Variola virus is the causative agent of small pox.Four immunogenic proteinsA27L and L1Rproteins specific to the intracellular mature virus(IMV)and the A33R and B5R proteins specific to the extracellular enveloped virus (EEV) are the best molecular candidates for the development of a small pox vaccine. The antigen was produced in soluble and insoluble forms upon transient and stable plant transformation.A27Land theA33Rproteinare stably expressed by Agrobacterium-mediated transformation of the nuclear genome and biolistic transformation of the plastome (Tiwari et al., 2009).

Patents regarding edible vaccine

S. No.	Patent holder	Claim
1	Prodigene	Recombinant antigen production and transfer to plants cells using plasmid vector system; Vaccine produced in genetically engineered plants for hepatitis and transmissible gastroenteritis virus
2	Found Advan Mil Med (USA)	Antibacterial vaccine expressed in plant cells, particularly useful against shigellosis
3	Ribozyme-Pharm	Nucleic acid vaccine used to treat or prevent viral infections in plants, animals or bacteria
4	Rubicon-Lab	Retrovirus expressed in animal or plant cells useful as virus and cancer vaccine
5	Applied Phytologics	Gene constructs for disease resistance, vaccine production in rice, barley, wheat, corn
6	Biosource (now Large Scale Biology)	Plant viral vector with potential as anti-AIDS vaccine; recombinant proteins for use in vaccines to protect against parasitic infection, eg malaria
7	University of Yale	Vaccine against invertebrates (insects, arachnids, helminthes, etc)
8	University of Texas	Hepatitis B virus core antigen recombinant vaccine
9	Biocem; Rhone-Merieux	Rabies vaccine in transgenic plants
10	Institute Pasteur	Attenuated *E coli* vaccine for use in gene therapy
11	University of Texas A&M/Tulane University	Transgenic plants containing *E coli* enterotoxin B for edible vaccine application in animals
12	USDA/Univ. Philadelphia	Rabies vaccine expressed in tomato plant
13	Scripps Research Institute	Recombinant antigen production in lettuce, spinach, tobacco, kidney bean, or *Chenopodium amaranticolor*
14	Cornell University	Increasing foreign protein expression
15	University Loma Linda	Gene constructs used to produce edible vaccines to treat autoimmune diseases
16	Agr Genet/ Purdue Research Foundation	Modified viruses used for vaccine production in plants, esp. against food and mouth disease, HIV and human rhino virus

Curtsey **-Misra et al., 2008**

Foreign proteins can be synthesized using microbial cell culture, animal cell culture, plant tissue culture, transgenic animals and transgenic plants, plants and plant cells are now considered as viable and competitive expression systems for large-scale protein production. The development of transgenic plants and plant viral vectors for foreign protein synthesis has been reviewed extensively. Using plant tissue culture approach, plant cells in differentiated or de differentiated states are grown axenically in nutrient medium in bioreactors under controlled conditions, with foreign protein harvested from either the biomass or culture liquid, or a combination of both. Although plant tissue culture may not be suitable for all applications of foreign protein synthesis, such as food-based production of edible vaccines of foreign proteins .Suspended plant cells have been used in several recent studies as a means of producing a variety of foreign proteins. These include recombinant antibodies and antibody fragments, enzymes such as β-glucuronidase invertase etc. The ability of plant cells and organs to propagate indefinitely in tissue culture without the need for sexual reproduction offers a solution to other problems relating to gene segregation and long-term transgene stability in agricultural crops .Stream processing, protein secretion and stability recovery and purification of proteins from plant biomass is an expensive and technically challenging business, requiring multiple separation steps such as precipitation, adsorption, chromatography and infiltration.

Cell suspension culture based expression

Transgenic plants require cultivation, harvesting extraction and if required purification of recombinant proteins to be used as vaccine antigens. Plants grown under field conditions often shows variable expression of recombinant protein(s).Hence, good manufacturing practices(GMP)and doze standardization may pose difficulties in the use of field grown plants as a source of plant-derived vaccine antigens. Further, linkage between farm and protein handling units may make the operational chain complex and add to costs. On the other hand, plant cell/tissue culture and hairy root culture could provide significant advantage in controlled production of Therapeutic proteins. These processes are independent of seasonal variations and enable continuous supply of the product. The continuous secretion and recovery of foreign cellular and culture medium can minimize the time and cost of process standardization, improve protein recovery, make the process more easily reproducible and reduce protein degradation during handling. The targeting of recombinant proteins with appropriate signal peptides for extracellular secretion can mimic the natural process in plants .Some proteins can be recovered easily from the secretion fluid or culture media .The addition of protein stabilizing agents in to the suspension culture medium can increase the accumulation of recombinant protein.(Tiwari et al., 2009)

The expression of recombinant proteins in suspension and hairy root culture offers promising potential for exploitation as large bioreactors. The two tobacco plant cell lines, Bright Yellow-2(BY-2) and N.tobaccum-1(NT-1) are utilized extensively for foreign protein production because of easy transformation and synchronous growth in liquid culture[43-52]

Continuous rhizo secretion of recombinant proteins is another promising strategy. Root biomass can be significantly increased by hairy root formation using Agrobacterium rhizogenes.

Stable expression of vaccine antigens in plants

Plant/tissue	Promoter	Pathogen/causingagent	Disease	Antigenicprotein	Reference
Potato/tuber ,callus	Mannopine synthase	V.cholerae and rotavirus	Cholera and Gastroenteritis	CTB–Rotavirus, enterotoxin protein(NSP4)	Arakawaetal. (2001)
Tobacco/leaf (chloroplast)	Plastid rRNA operon(Prrn)	V.cholerae	Cholera	CTB	Danielletal. (2001a)
Potatotuber	Mannopine synthase	V. cholerae ,rotavirus and E.coli	Cholera, Gastroenteritis and Diarrhea	CTB–NSP4/CTA2,CFA/1	YuandLangridge (2001)
Tomato/fruit, leaf	CaMV35S	V.cholerae	Cholera	CTB	Janietal.(2002)
Potato/tuber	Mannopines ynthase	V.cholerae and rotavirus	Cholera and gastroenteritis	CTB–NSP4	Kim and Langridge (2003)
Tobacco/leaf	CaMV35S	V.cholerae	Cholera	CTB	Janietal.(2004)
Tobacco/leaf	CaMV35S	V.cholerae	Cholera	CTB–InsB3	Lietal.(2006b)
Tobacco/leaf	CaMV35S	V.cholerae	Cholera	CTB	Mishraetal. (2006)
Tobacco/hairyroot	CaMV35S	V.choleraeand	Cholera	CTB–Surfaceprotectiveantigen(SpaA)	Koe tal.(2006)

Plant/tissue	Promoter	Pathogen	Disease	Antigen	Reference
lettuce/leaf	CaMV35S	Erysipelothrix Rhusiopathiae V.cholerae	Cholera	sCTB–KDEL	Kimetal.(2006)
Tobacco/leaf	PsbA	V.cholerae and Canine parovirus (CPV)	Choleraand Haemorrhagic Gastroenteritisand myocarditis	CTB–2L21	Molinaetal. (2004)
Tomato/fruitandtobacco/leaf,flower,seed	FruitspecificE8and	V.choleraandHepatitisB	CholeraandHepatitis	CTBandHBsAg	Heetal.(2008)
Tomato/fruit	CaMV35Swithleader Peptide from alfalfa mosaic	V.cholerae	Cholera	CTB-P4/CTB-P6/TCPA	Sharmaetal. (2008)
Potato/tuber,tobacco/leaf	CaMV35S	E. coli	Diarrhea	HeatlabiletoxinBsubunit(LTB)	Haqetal.(1995)
1. Potato/tuber	Tuber-specificpatatin	E. coli	Diarrhea	LTB	Lauterslager etal.(2001)
Maize/seed	CaMV35S	E.coliandSwinetransmissiblegastroenteritiscoronavirus(TGEV)	DiarrheaandSwinetransmissiblegastroenteritis(TGE)	LTBandTGEVglycoproteinS	Streatfieldetal., 2001;Lamphearetal.,2002
Maize/seed	Endospermspecific gammazein	E. coli	Diarrhea	LTB	Chikwambaetal.(2002)
Tobacco/leaf	CaMV35 S	EnteropathogenicE.coli (EPEC)	Diarrhea	Bundle-forming pilus structural subunit A(BfpA)	daSilva etal. (2002)
Tomato/fruit,leaf	CaMV35S	E. coli	Diarrhea	LTB—Mousezonapellucida3epitope(ZP3)	Walmsley etal. (2003)
Arabidopsis thaliana/leaf	CaMV35S	E. coli-M. tuberculosis, M. Bovis	Diarrheaand Tuberculosis	LTB—Earlysecretoryantigenictarget-6 (ESAT-6)	Riganoetal. (2004)
Tobacco/leaf	CaMV35S	E. coli	Diarrhea	LTB—SEKDEL	Kang etal.(2005)
Siberian ginseng/ somaticembryos	CaMV35SandUbiquitin	E. coli	Diarrhea	LTB	Kang etal. (2006b)
Soybean/seed	Seedspecificglycinin	E. coli	Diarrhea	LTB	Moravecetal. (2007)
Carrot/leaf,root	CaMV35S	E. coli	Diarrhea	LTB	Rosales-Mendoza etal. (2008)
Tobacco/chloroplast	plastid16S rRNA gene promoter(Prrn)	Enterotoxigenic E.coli (ETEC)strains	Diarrhea	LTB—Heatstabletoxin(ST)	Rosales-Mendoza etal. (2009)
Potato/tuber	CaMV35S	Porcine epidemic diarrheavirus(PEDV)			
Tobacco/leaf	CaMV35S	Porcine epidemic diarrheavirus(PEDV)			
Potato/tuber,potato/leaf	Tuber specific patatin and CaMV35S	HBV	Hepatitis	HBsAgandHBsAg-VSPαS/VSPαL	Richter etal. (2000)
Potato/tuber	CaMV35S	HBV	Hepatitis	HBsAg	Kong etal.(2001)
TobaccoNT1andsoybean	CaMV35S	HBV	Hepatitis	HBsAg	Smithetal.(2002a,b)
W82cell suspension cultures	CaMV35S	HBV	Hepatitis	HBs	
Cherrytomatillo/ stem, fruitTobacco/NT-1celllineculture	CaMV35S	HBV	Hepatitis	HBsAg-VSPαS	Sojikul etal. (2003)
Potato/tuber	TuberspecificpatatinandCaMV35S	HBV	Hepatitis	HBsAgSandpreS2antigens	Joung etal. (2004)
Tobacco/leafandtomato/leaf,fruit	CaMV35S	HBV	Hepatitisand Gastroenteritis	HBsAgM/SandNVCP	Huanget al.(2005)
Banana/fruit,leaf	ubq3andEthyleneforming enzyme(EFE)	HBV	Hepatitis	HBsAg	Kumaretal. (2005a)
Tobacco/cell linesuspensionculture	ubq3andEFE	HBV	Hepatitis	HBsAg	Kumaretal. (2005b)

Plant/tissue	Promoter	Pathogen	Disease	Antigen	Reference
Potato/tuber,hairy root	EFE	HBV	Hepatitis	HBsAg	Kumaretal.(2006)
Potato/tuber	CaMV35S	HBV	Hepatitis	HBsAgM	Youmaetal.(2007)
Tomato/fruit	Fruitspecific2A11	HBV	Hepatitis	PRS-S1S2S	Louetal.(2007)
Potato/tuber,tobacco/leaf	Tuber specific patatin and CaMV35S	Norwalkvirus(NV)	Gastroenteritis	Nor walk virus capsid protein(NVCP)	Masonetal.(1996)
Potato/leaf, tuber	CaMV35S	BovinegroupArotavirus (GAR)	Severe viral diarrheain humans and animals	MajorcapsidproteinVP6	Matsumuraet al.(2002)
Potato/tuberleaf	CaMV35S	Rotavirus	Gastroenteritis	capsid ofrotavirusglycoproteinVP7	Wuetal.(2003)
Alfalfa/leaf	CaMV35S	Bovinerotavirus(BRV)	Gastroenteritisin mammals		Wigdorovitz etal.(2004)
Alfalfa/leaf	CaMV35S	Rotavirus	Viral gastroenteritis	PBsVP6 humangroupArotavirus	Dongetal.(2005)
Potato/tuber,leaf	CaMV35S	Rotavirus	Gastroenteritis	capsid ofrotavirusglycoproteinVP7	Lietal.(2006a)
Potato/tuber,leaf	CaMV35S	humanpapillomaviruses (HPV)	Cervicalcancer	HPV11 L1majorcapsidprotein	Warzechaet al.(2003)
Tobacco/leaf,potato/tuber, leaf	CaMV35S	humanpapillomaviruses (HPV)	Cervicalcancer	HPV16 Virus-LikeParticles	Biemeltet al.(2003)
A.thaliana/leaf,tobacco/ leaf	CaMV35S	HPV	Cervicalcancer	HPV11 L1majorcapsidprotein	Kohletal.(2007)
Tomato/leaf,fruitCaMV35SCorynebacteriumdiphteriae,BordetellaPertussisandClostridiumte tani			Diphtheria,PertussisandTetanus(DPT)	epitopesoftheC.diphtheriae,B.pertussis andC.tetaniexotoxins	Soria-Guerra et al.(2007)
Tomato/leaf,fruit	CaMV35S	Rabiesvirus	Rabies	RGP	McGarveyetal.(1995)
Tobacco/leaf	CaMV35S	Rabiesvirus	Rabies	RGP	Ashrafetal.(2005)
Tomato/leaf,fruit	CaMV35S	Rabiesvirus	Rabies	Rabiesnucleoprotein (RNP)	Arangoetal.(2008)
Rice/leaves,seeds	ubiquitinandseedspecific glutelin	Newcastlediseasevirus (NDV)	Newcastledisease (ND)	NDVenvelop	
Tobacco/leaf	CaMV35S	Measlesvirus (paramyxovirus)	Measles	MV-H (Measelsvirushemagglutinin)	Hung n etal.(2001)
Tobacco/leaf	CaMV35S	Measelsvirus	Measles	MV-H	Web steretal.(2002)
Carrot/leaf,root	CaMV35S	Measlesvirus	Measles	MV-H	Marquet-Blouinetal.(2003)
Tobacco/leaf	CaMV35S	Measelsvirus	Measels	MV-H	Web steretal.(2005)
Peanut/leaf	CaMV35S	Rinderpestvirus(RPV)	Rinderpest	Hproteinofrinderpestvirus	Khandelwaletal.(2003)
Collard/leaf cauliflower/ floret of mature curd	CaMV35Sandsynthetic OCS3MAS.	Vacciniavirus,human SARScoronavirus	Smallpox and human SARS	vacciniavirusB5coatprotein and corona virus spike glycoproteinepitope	Pogrebnyaketal.(2006)
			Diarrhea	Neutralizing epitope of PEDV(COE)	Kimet al.,(2005)
			Diarrhea	Neutralizing epitope of PEDV(CO-26K)	Kang etal.(2006a

Potato/leaf, tuber	CaMV35S	Bovine group A rotavirus (GAR)	Severe viral diarrhea in humans and animals	Major capsid protein VP6	Matsumura et al.(2002)
Potato/tuber, leaf	CaMV35S	Rotavirus	Gastroenteritis	capsid of rotavirus glycoprotein VP7	Wu et al.(2003)
Alfalfa/leaf	CaMV35S	Bovine rotavirus (BRV)	Gastroenteritis in mammals	eBRV4	Wigdorovitz et al.(2004)
Alfalfa/leaf	CaMV35S	Rotavirus	Viral gastroenteritis	PBsVP6	human group A rotavirus Dong et al.(2005)
Potato/tuber, leaf	CaMV35S	Rotavirus	Gastroenteritis	capsid of rotavirus glycoprotein VP7	Li et al.(2006a)
Potato/tuber, leaf (HPV)	CaMV35S	human papillomaviruses	Cervical cancer	HPV11 L1 major capsid protein	Warzecha et al.(2003)
Tobacco/leaf, potato/tuber, leaf	CaMV35S	human papillomaviruses (HPV)	Cervical cancer	HPV16 Virus-Like Particles	Biemelt et al.(2003)
A.thaliana/leaf, tobacco/leaf	CaMV35S	HPV	Cervical cancer	HPV11 L1 major capsid protein	Kohl et al.(2007)
Tomato/leaf, fruit	CaMV35S	Corynebacterium diphtheriae, Bordetella Pertussis and Clostridium tetani	Diphtheria, Pertussis and Tetanus (DPT)	epitopes of the C.diphtheriae, B.pertussis and C.tetani exotoxins	Soria-Guerra et al.(2007)
Tobacco/leaf	CaMV35S	Rabies virus	Rabies	RGP	Ashraf et al.(2005)
Tomato/leaf, fruit	CaMV35S	Rabies virus	Rabies	Rabies nucleoprotein (RNP)	Arango et al.(2008)
Rice/leaves, seeds	ubiquitin and seed specific glutelin	Newcastle disease virus (NDV)	Newcastle disease (ND)	NDV envelope fusion (F) glycoprotein	Yang et al.(2007)
Tobacco/leaf (paramyxovirus)	CaMV35S	Measles virus	Measles	MV-H (Measels virus hemagglutinin)	Huang et al.(2001)
Tobacco/leaf	CaMV35S and synthetic OCS3MAS.	Vaccinia virus, human SARS coronavirus	Smallpox and human SARS	vaccinia virus B5 coat protein and coronavirus spike glycoprotein epitope	Pogrebnyak et al. (2002) / Wasser et al. (2002)

Carrot/leaf,root	CaMV35S	Measlesvirus	Measles	MV-H	Marquet-Blouinetal.(2003)
Tobacco/leaf	CaMV35S	Measelsvirus	Measels	MV-H	Websteretal.(2005)
Peanut/leaf	CaMV35S	Rinderpestvirus(RPV)	Rinderpest	Hproteinofrinderpestvirus	Khandelwaletal.(2003)
	CaMV35Sandsynthetic OCS3MAS.	Vacciniavirus,human SARScoronavirus	Smallpoxandhuman SARS	vacciniavirusB5coatprotein and coronavirusspikeglycoproteinepitope	Pogrebnya
Arabidopsis/leaf	CaMV35S Swinetransmissiblegastroenteritiscorona virus (TGEV)		Transmissible gastroenteritis(TGE)	GlycoproteinS	Gómezet al.(1998)
Potato/tuber	CaMV35S	TGEV	TGE	GlycoproteinS	Gómezetal.(2000)
Tobacco/leaf	SyntheticsuperpromoterTGEV		TGE	GlycoproteinS	Tubolyetal.(2000)
Arabidopsis/leaf	CaMV35S	Footandmouth diseasevirus(FMDV)	Footandmouthdi sease	StructuralproteinVP1	Carrilloet al.(1998)
Potato/tuber,leaf	CaMV35S	FMDV	Footandm outhdisease	StructuralproteinVP1	Carrilloet al.(2001)
Alfalfa/leaf disease	CaMV35S	FMDV	Footandmouth	StructuralproteinVP1	DusSantosetal.(2002)
Potato/leaf,tuber	CaMV35S	Rabbithaem orrhagicdiseasevirus(RHDV)	Rabbithemorrhagics yndrome	StructuralproteinVP60	Castañónet al.(1999)
Potato/leaf,tuber	CaMV35S,sunflower PolyubiquitiandB33p atatin	RHDV	Rabbithem orrhagicsyndrome	StructuralproteinVP60	Castañónet al.(2002) Martín-Alonso etzal.(2003)Curtissand
Arabidopsis/leaf	CaMV35S	Canineparvovirus	Canineparvo virusdisease	VP2capsidproteinofcanineparvovirus (CPV)	Gilet al.(2001)
White	clover/leaf CaMV35SMann heimiahaemolytica(bovin epneumoniapasteurellos is)		Bovineviraldiarrhea	Leukotoxin(Lkt)	Leeetal.(2001)
Tobacco/leaf	CaMV35S	Bacillus anthracis	Anthrax	Protectiveantigen	Azizetal.(2002)
Tobacco/leaf(chloroplast)	PlastidrRNAoperon(Prrn)	Clostridiumtetani	Tetanus	Tetanusvaccine antigen(TetC)	Tregoningetal.(2003)
Tobacco,collard/leaf	rbcSandCaMV35S	Vacciniavirus	Smallpox	VacciniavirusB5coatprotein	Golovkinetal.(2007)
Tomato/fruit	FruitspecificE8	Human immunodeficiencyvirus(HIV)	AIDS	TatproteinofHIV-1	Ramfrezetal.(2007)
Potato/tuber	CaMV35S	Causedbyneuro-degeneration	Alzheimer's	Humanβ-amyloid(Aβ)	Kimetal.(2003)
Tomato/leaf	CaMV35S	Causedbyneuro-degeneration	Alzheimer's (2008)	Humanβ-amyloid(Aβ)	Youmaetal.
Rice/seeds jointcartilage	Seed specificglutelinA	Causedbyinflammationof	Arthritis	Type IIcollagenpeptide	Hashizumeet al.(2008)
Peanut/callus	CaMV35S	Bluetonguevirus(BTV)	Bluetongue	VP2genethatconstitutestheoutercapsid ofthevirus	Athmarametal.(2006)
Papaya/Embryogenicclones (ETgpC)	CaMV35S	Taeniasolium	Cysticercosis	SyntheticpeptidesKETc1,KETc12,KETc7	Hernándezetal.(2007)
Tomato/leaf,fruit	CaMV35S	Organophosphate poisoning	Severeacutepancr eatitisandmyocar dialinjury	Humanacetylcholinesterase(AchE)	Moretal.(2001)
Tomato/fruit	CaMV35Sandfruitspecific E8	Respiratorysyncytialvirus (RSV)	Seriousrespiratoryt ractdisease	RSVFprotein (2000)	Sandhuetal.
Tobacco/leaf,lettuce/leaf	CaMV35S	SevereAcuteRespiratory		SyndromeCoronavirus	

Plant/tissue	Promoter	Pathogen	Disease	Antigen	Reference
			Severe Acute Respiratory Syndrome	Partial spike(S) protein of SARS-CoV	Li et al.(2006c)
Tobacco/leaf	CaMV35S	Avian influenza virus H5/HA1 variant	Avian flu	H5/HA1 variant-HDEL	Spitsin et al.(2009)
Tomato/fruit	CaMV35S	Yersinia pestis	Pneumonic/bubonic plague	Antiphagocytic capsular envelope glycoprotein(F1) and low calcium response virulent antigen (V) fusion protein	Alvarez et al.(2006)
Tobacco/chloroplast	psbA	Yersinia pestis	Pneumonic/bubonic plague	F1-V fusion protein	Arle net al.(2008)
Tobacco and Arabidopsis/leaf	CaMV35S	Human immunodeficiency virus (HIV-1) and hepatitis B virus (HBV)			

courtesy Tiwari et al, 2009

Conclusion

Plant derived vaccines have a great future as a cost-effective-to-administer, easy-to-store, readily acceptable vaccine delivery system, especially for the poor developing countries. It involves introduction of selected desired genes into plants and then inducing these altered plants to manufacture the encoded proteins. This was a mere imagination now it becomes a reality today. A variety of delivery systems have been developed. Initially thought to be useful only for preventing infectious diseases, it has also found application in prevention of autoimmune diseases, birth control, cancer therapy, etc. These vaccines are currently being developed for a number of human and animal diseases. There is growing acceptance of transgenic crops in both industrial and developing countries. Resistance to genetically modified foods may affect the future of edible vaccines. They have passed the major hurdles in the path of an emerging vaccine technology. Various technical obstacles, regulatory and non-scientific challenges, though all seem surmountable, need to be overcome. This review attempts to discuss the current status and future of this new preventive modality. (Lal et al .2007)

Edible Vaccines have also crossed the human barrier to leap into animal vaccines which are further extended into even developing countries. These systems though still face some hurdles in the form of technical, regulative, legislative challenges etc which will need to be overcome to sustain the future of these vaccine systems. Analyzing Edible Vaccines is an in depth study on the basics of vaccines and the major types of vaccines which are present in the traditional vaccine delivery systems. The report presents in detail the theoretical basics of edible vaccines and a close understanding of the action mechanism of the edible vaccines. Further, the report elaborates on the preparation of edible vaccines and the advances which have led to the achieving the second generation edible vaccines. As every technology faces its challenges the documented report emulates how the controversy over genetically modified food products, regulatory issues and other challenges are affecting the faster growth of edible vaccine systems.[112]

References

1. Arntzen C. Plant-derived vaccines and antibodies potential and limitations.Vaccine2005, 23:1753–6.

2. www.care.org/campaigns/children poverty facts.asp

3. Larrick W., Thomas D.W. Producing proteins in transgenic plants and animals .Curr Opin Biotechnol2001,12:411–8.

4. Houdebine L. M. Production of pharmaceutical proteins by transgenic animals. Comp Immunol Microbiol Infect Dis.2009, 32:107–21.

5 Ma J. K.-C, Drake PMW, Christou P. The production of recombinant pharmaceutical proteins in plants. Nat Rev Genet 2003; 4:794–805

6. Koprowski H. Vaccines and sera through plant biotechnology.Vaccine2005,23,1757–63.

7. Lal P., Ramachandran V. G., Goyal R., Sharma R. Edible vaccines: current status and future. Ind J Med Microbiol 2007;25:93-102

8.Mishra N., Gupta P. N., Khatri K., Goyal A. K., Vyas S. P. Edible vaccines :a new approach to oral immunization. Ind. J. Biotechnol 2008,7,283–94.

9Sijmons P. C., Dekker B.M., Schrammeijer B., Verwoerd T.C., van den Elzen P.J., Hoekema A. Production of correctly processed human serum albumin in transgenic plants. Biotechnology (NY)1990,8: 217–221.

10. Twyman R. M., Schillberg S., Fischer R. Transgenic plants in the biopharmaceutical market. Expert Opin Emerg Drugs 2005,10: 185–218.

11. Twyman R.M., Stoger E., Schillberg S., Christou P., Fischer R. Molecular farming in plants: host systems and expression technology. Trends Biotechnol 2003, 21: 570–578.

12. Subbirhussain Edible Vaccines – modern approach in plant biotechBy Technology Times ,2011, June 25,

13. Menkhaus T.J., Bai Y., Zhang C., Nikolov Z.L., Glatz C.E. Considerations for the recovery of recombinant proteins from plants. BiotechnolProg,2004, 20: 1001–1014.

14. Chen M., Liu X., Wang Z., SongJ., Qi Q, Wang P. G. Modification of plant N- glycans processing: the future of producing therapeutic protein by transgenic plants. Med Res Rev2005;25:343–60

15. Yonekura –Sakakibara K., Saito K. Genetically modified plants for the promotion on human health.BiotechnolLett2006,28,1983–91

16.Hood E. E., Woodard S. L., Horn M. E. Monoclonal antibody manufacturing in transgenic plants—myths and realities.Curr.Opin.Biotechnol.2002,13,630–5

17. Stevens L. H., Stoopen G.M., Elbers I.J., Molth off J. W., Bakker H. A., Lommen A. Effect of climate conditions and plant developmental stage on the stability of antibodies expressed in transgenic tobacco. Plant Physiol 2000,124,173–82.

18. Stoger E., Ma J. K-C, Fischer R., Christou P. Sowing the seeds of success: pharmaceutical proteinsfromplants. Curr Opin. Biotechnol. 2005;16:167–73.

19.Benchabane M., Goulet C., Rivard D., Faye L., Gomord V., Michaud D. Preventing unintended proteolysis in plant protein biofactories. Plant Biotechnol J.2008,6,633–48

20. Daniel H., Streat field S. J., Wycoft K. Medical molecular farming production of antibodies, biopharmaceuticals and edible vaccines in plants. Trends Plant Sci 2001b,6,219–26.

21. Chen M., Liu X., Wang Z., Song J,Qi Q, Wang P.G. Modification of plant N-glycans processing: the future of producing therapeutic protein by transgenic plants.Med ResRev 2005;25:343–60

22. Stoger E. ,Ma JK-C, Fischer R.,Christou P.Sowing the seeds of success: pharmaceutical proteinsfrom plants.Curr Opin Biotechnol 2005,16,167–73.

23. Muntz K. Deposition of storage proteins.PlantMolBiol 1998;38:77–99.

24. Kumar G. B. S.,Ganapathi T.R., Revathi C. J.,Srinivas L.,Bapat V.A. Expression of hepatitisB Surface antigen intransgenic banana plants. Planta 2005a;222:484–93.

25.KumarG.B.S.,GanapathiT.R.,SrinivasL.,RevathiC.J.,BapatV.A.Secretion of hepatitis B surface antigen intransformed tobacco cell suspension cultures. Biotechnol Lett2005b;27:927–32.

26.Benchabane M., Goulet C., Rivard D.,Faye L., Gomord V.,Michaud D. Preventing unintended proteolysis in plant protein bio factories.Plant Biotechnol J2008; 6:633–48.

27.Fischer R., Stoger E., Schillberg S., Christou P., Twyman R. M. .Plant-based production of bio pharmaceuticals. Curr Opin PlantBiol2004; 7:152–8.

28. The use of opened multi dose vials of vaccinesin subsequent immunization sessions WHO Policy.Dep. of Vaccines and Biologists World Health Organisation,Geneva 2000.

29. Mason H.S., Lam D.M., Arntzen C.J. Expression of hepatitis B surface antigen in transgenic plants. Proc Natl Acad Sci USA 1992, 89 , 11745-9.

30. Mason H.S., Haq T.A, Clements J.D., Arntzen C.J. Edible vaccine protects mice against *Escherichia coli* heat-labile enterotoxin (LT): Potatoes expressing a synthetic LT-B gene. Vaccine 1998, 16 ,1336-43.

31. Wongsamuth R., Doran P.M. Production of monoclonal anti bodies by tobacco hairy roots. Biotechnol Bioeng 1997, 5.

32. Spitsin S., Andrianov V., Pogrebnyak N., Smirnov Y., Borisjuk N, Portocarrero C,etal. Immunological assessment of plant-derived avian flu H5/HA1variants.Vaccine2009,27 : 1289–92.

33.Alvarez M. L., Pinyerd H. L., Crisantes J. D., Rigano M. M., Pinkhasov J., Walmsley A. M. Plant-made sub unit vaccine against pneumonic and bubonic plague is orally immunogenicinmice.Vaccine2006,24,2477

34. Greco R., Michel M., Guetard D., Cervantes-Gonzalez M., Pelucchi N., Wain-Hobson S. Production of recombinant HIV-1/HB V virus-like particles in *Nicotiana tabacum* and *Arabidopsis thaliana* plants for a bivalent plant-basedvaccine.Vaccine2007,25, 8228–40

35. William H. R. edible vaccine Langridge Scientific American September 2000

37. Rigano M. M., Manna C., Giulini A., Vitale A., Cardi T. Plants as bio factories for the production of sub unit vaccines against bio-security-related bacteria and viruses.

Vaccine2009.

38. Moravec T., Schmidt M. A. , Herman E. M., Woodford- Thomas T. Production of Escherichia coli heat labile toxin (LT)B subunit in soybean seed and analysis of its immunogenicity as an oral vaccineVaccine2007,25,1647–57.

39.Rosales-MendozaS.,Alpuche-SolısA.G.,Soria-GuerraR.E.,Moreno-FierrosL.,Martnez Gonzalez L., Herrera Dıaz A,. Expression of an Escherichia coli antigenic fusion protein comprising the heat labile toxin B subunit and the heat stable toxin, and its assembly as a functional oligomer in transplastomic tobacco plants. Plant J.2009,57, 45–54.

40. Richter L.J., Thanavala Y., Arntzen C.J., Mason H.S. Production of hepatitis B surface antigen in transgenic plants for oral immunization. Nat Biotechnol .2000,18, 1167–1171.

41Carrillo, C., Wigdorovitz, A., Oliveros, J. C., Zamorano, P. I., Sadir, A. M., Gomez, N., Salinas, J., Escribano, J. M. and Borca, M. V., J. Virol., 1998, 72, 1688–1690.

42. D. E. Webster, L. Wang, M. Mulcair "Production and characterization of an orally immunogenic Plasmodium antigen in plants using a virus-based expression system," Plant Biotechnology Journal,2009, vol. 7, no. 9, pp. 846–855.

43. La Count W., An G., Lee J.M. The effect of poly vinyl pyrrolidone (PVP)on the heavy chain monoclonal antibody production from plant suspension cultures. Biotechnol Lett 1997, 19:93-96.

44. Wongsamuth R., Doran P.M. Production of monoclonal anti bodies by tobacco hairy roots. Biotechnol Bioeng 1997, 5.

45. Sharp J. M., Doran P. M. Effect of bacitracin on growth and monoclonal antibody production by tobacco hairy roots and cell suspensions. Biotechnol Bioprocess Eng 1999, 4,253-258.

46. Fischer R., Schumann D., Zimmermann S., Drossard J., Sack M., Schillberg S. Expression and characterization of bi specific single chain Fv fragments produced in transgenic plants. Eur J Biochem1999, 262,810-816.

47. Fischer R., Liao Y-C., Drossard J.Affinity-purification of a TMV specific recombinant full-size antibody from a transgenic tobaccosuspension culture. J Immunol Methods 1999, 226,1-10.

48. Fischer R., Emans N., Schuster F., Hellwig S., Drossard J. Towards molecular farming in the future: using plant-cell-suspension cultures as bioreactors. Biotechnol Appl Biochem 1999, 30,109-112.

49. Liu F., Lee J.M. Effect of culture conditions on monoclonal antibodyproduction from genetically modified tobacco suspension cultures. Biotechnol Bioprocess Eng 1999, 4,259-263.

50. Kurata H., Takemura T., Furusaki S., Kado C.I. Light-controlled expression of a foreign gene using the chalcone synthase promoter in tobacco BY-2 cells. J Ferment Bioeng 1998, 86,317-323.

51. Verdelhan des Molles D., Gomord V.,Bastin M., Faye L. Courtois D.Expression of a carrot invertase gene in tobacco suspension cellscultivated in batch and continuous culture conditions. J Biosci Bioeng 1999, 87,302-306.

52. Magnuson N.S., Linzmaier P.M., Reeves R., An G., Hay Glass K., Lee J.M. Secretion of biologically active human interleukin-2 andinterleukin-4 from genetically modified tobacco cells in suspension culture. Protein Expr Purif 1998, 13,45-52.

53.SiddharthTiwari,PraveenC.Verma,PradhyumnaK.Singh, RakeshTuli Plants as bioreactors for the production of vaccine antigens BiotechnologyAdvances2009,27, 449–4

54 Araka.waT., Yu J., Langridge W. H. Synthesis of a choleratoxin B subunit-rotavirus NSP4 fusion protein in potato.PlantCellRep 2001,20,343–8.

55.Daniell H., Lee S. B., Panchal T. ,Wiebe P. O. Expression of the native cholera toxin B subunit gene and assembly as functional oligomers.J Mol Biol 2001a,311,1001–9.

56. Yu J., Langridge W.H.A. plant based multi component vaccine protects mice from enteric diseases. Nat Biotechnol 2001,19,548–52.

57Jani D., Singh N.K.,Bhattacharya S, Meena L.S.,SinghY.,Upadhyay S.N.,.Studies on the immunogenic potential of plant-expressed cholera toxin B subunit .Plant Cell Rep2004,22,471–7

58. Kim T.G., Langridge W. H. R. Assembly of cholera toxin B subunit full-length rotavirus NSI fusion protein oligomers in transgenic potato. Plant Cell Rep 2003,21,884–90

59.Jani D., Singh N.K., Bhattacharya S., Meena L.S.,Singh .Y,Upadhyay S.N., Studies on the immunogenic potential of plant-expressed cholera toxin B subunit. Plant Cell Rep 2004,22, 471–7.

60.LiD,O'Leary J., Huang Y., Huner N. P. A. , Jevnikar A.M., Ma S. Expression of cholera to subunit and the B chain of human insulin as a fusion protein in transgenic tobacco plants. Cell Rep 2006b,25,417–24.

61.Mishra S.,Yadav D. K.,Tuli R..Ubiquitin fusion enhances cholera toxin B subunit expressio In transgenic plants and the plant-expressed protein binds GM1 receptors more efficiently.. Biotechnol 2006,127,95-108.

62.Ko S., Liu J. R.,Yamakawa T., Matsumoto Y. Expression of the antigen (SpaA)in transgenic hairy roots of tobacco. Plant Mol Biol Report 2006,24,251a–g.

63.KimY. S., Kim B. G.,KimT.G.,Kang T. J., Yang M. S. Expression of a cholera toxin B subunit intransgenic lettuce(LactucasativaL.) using Agrobacterium mediated transformation system. Plant Cell Tissue Organ Cult 2006,87,203–10.

64.Jiang X.L.,He Z.M.,Peng Z.Q.,QiY.,Chen Q.,Yu S.Y.Choleratoxin B protein in transgenic tomato fruit induces systemic immune response in mice. Transgenic Res2007,16,169– 75.

65.NochiT.,Takagi H.,Yuki Y.,Yang L.,MasumuraT.,Mejima M., Rice-based mucosal vaccinea a global strategy for cold-chain-and needle free vaccination. ProcNatlAcadSci USA 2007, 104,10986–91.

66. He Z. M., Jiang X. L., Qi Y.,Luo D.Q.Assessment of the utility of the tomato fruit-specific E8 promoter for driving vaccine antigen expression.Genetica2008,133,207–14.

67.Sharma M. K.,Singh N. K., Jani D.,Sisodia R., Thungapathra M.,Gautam J. K .Expression of toxinco-regulated pilus sub unit A(TCPA) of Vibrio cholera and its immunogenic epitopes fused to cholera toxin B sub unit intransgenic tomato (*Solanum lycopersicum*). PlantCellRep 2008,27,307–18.

68.Oszvald M.,Kang T. J. ,Tomoskozi S. ,Jenes B. ,Kim T. G., Cha Y. S, Expression of cholera toxin B sub unit intransgenic rice endosperm.Mol Biotechnol 2008,40,261–8.

69. Lauterslager T. G. M. ,Florack D. E. A. ,Vander Wal T. J.,Molthoff J. W., Langeveld J. P. M., Bosch D, Oral immunization of naïve and prime danimals with transgenic potato tubers expressingLT-B.Vaccine 2001,19,2749–55.

70.Streatfield S. J.,Howard J. A .Plant-based vaccines.Int J Parasitol2003a,33,479–93.

71. Chikwamba R., Cunnick J., Hathaway D., Mc Murray J.,Mason H.,Wang K.A. functional antigen in a practical crop: LT-B producing maize protects mice against Escherichia coli heat labile enterotoxin (LT)and cholera toxin (CT). TransgenicRes2002,11,479–93

72. Da Silva J. V., Garcia A. B.,Flores V.M.Q., Macedo Z. S., Medina-Acosta E. Phyto secretion of entero pathogenic Escherichia coli pilin subunit A intransgenic tobacco and its suitability for early life vaccinology.Vaccine2002,20,2091–101.

73. Moravec T. ,Schmidt M. A., Herman E.M.,Woodford-ThomasT. Production of Escherichia coli heat labile toxin (LT)B subunit in soybean seed and analysis of its immunogenicity as an oral vaccine.Vaccine2007,25,1647–57.

74.Rosales-Mendoza   S.,Alpuche-SolisA.G.,Soria-GuerraR.E.,Moreno-Fierros   L.,Martnez-Gonzalez L.,Herrera-Diaz A., Expression of an Escherichia coli antigenic fusion protein comprising the heat labile toxin B subunit and the heat stable toxin, and its assembly as a functional oligomer in transplastomic tobacco plants. Plant J 2009,57, 45–54.

75.Rosales-MendozaS,Alpuche-SolisAG,Soria-GuerraRE,Moreno-FierrosL,Martnez-Gonzalez L, Herrera-DiazA, Expression of an Escherichia coli antigenic fusion protein comprising the heat labile toxin B sub unit and the heat stable toxin, and its assembly as a functional oligomer in transplastomic tobacco plants. Plant J2009;57:45–54.

76. Richter L. J., Thanavala Y., Arntzen C. J., Mason H. S. Production of hepatitis B surface anti gen in transgenic plants for oral immunization .Nat Biotechnol2000,18,1167–71.

77. Kong Q.,Richter L., Yang Y.F., Arntzen C.J., Mason H.S.,ThanavalaY. Oral immunization with hepatitis surface antigen expressed in transgenic plants. Proc Natl Acad Sci USA 2001,98,11539–44.

78. Smith M. L., Keegan M. E., Mason H. S., Shuler M. L. Factors important in the extraction, stability and in vitro assembly of the hepatitis surface antigen derived from recombinant plant systems .Biotechnol Prog. 2002a,18, 538–50.

79.Smith M. L.,Mason H.S.,Shuler M.L..Hepatitis B. surface antigen (HbsAg)expression nin plant cell culture: kinetics of antigen accumulation in batch culture and its intracellular form. Biotechnol Bioeng2002b,80,812–2

80.JiangX. L, He Z.M.,PengZ.Q.,QiY.,Chen Q.,Yu S.Y. Cholera toxin B protein in transgenic tomato fruit induces systemic immune response in mice. TransgenicRes 2007, 16,169–75.

81.Huang Z.,Elkin G.,Maloney B.J.,Beuhner N.,Arntzen C.J.,ThanavalaY., Virus-like particle expression and assembly in plants hepatitisB and Norwalk viruses. Vaccine2005,23,1851–8.

82.Kumar G.B.S.,GanapathiT.R.,RevathiC.J.,SrinivasL.,BapatV.A.Expression of hepatitis B Surface antigen in transgenic banana plants.Planta2005a,222,484–93.

83.Kumar G.B.S.,GanapathiT.R.,Srinivas L.,Revathi C.J., Bapat V.A. Secretion of hepatitis surface antigen in transformed tobacco cell suspension cultures. Biotechnol Lett2005b,27,927–32.–74

84.Youma J. W., Won Y. S., Jeon J. H., Ryu C. J., Choi Y. K., Kim H. C. Oral immunogenicity of potato-derived HBs Ag middle protein in BALB/cmice.Vaccine2007,25,577–84.

85.Lou X.M., Yao Q.H., Zhang Z., Peng R. H., Xiong A. S., Wang H. K. Expression of the human hepatitis virus large surface antigen gene intransgenic tomato plants.ClinVaccine Immunol,2007,14,464–9.

90.MatsumuraT.Itchoda N,Tsunemitsu H. Production of immunogenic VP6 protein of bovine group A rota virus intransgenic potato plants.ArchVirol 2002,147,1263–70.

91.Wu Y. Z.,Li J. T.,Mou Z. R.,FeiL,Ni. B.,Geng M., Oral immunization with rota virus VP7 expressed in transgenic potatoes induced high titers of mucos alneutralizing IgA. Virology,2003,313,337–42

92.Wigdorovitz A., Mozgovoj M., DusSantos M. J., Parreño V., Gómez C., Pérez-Filgueira D.M..Protective lactogenic immunity conferred by an edible peptide vaccine to bovine rotavirus produced intransgenic plants.J GenVirol2004,85,1825–32.

93Dong J. L., Liang B. G., Jin Y. S., Zhang W. J., Wang T. Oral immunization with p BsVP6-transgenic alfalfa protects mice against rotavirus in fection.Virology2005, 339, 153–63.

.94.Li J. T, Fei L., Mou Z. R.,Wei J., Tang Y., He H. Y.,. Immunogenicity of a plant- derived edible rotavirus subunit vaccine transformed over fifty generations.Virology2006a, 356, 171–8.

95. Wigdorovitz A., Mozgovoj M., DusSantos M. J., ParreñoV.,Gómez C.,Pérez-Filgueira D. M. Protective lactogenic immunity conferred by an edible peptide vaccine to bovine rota virus produced in transgenic plants.J GenVirol2004,85,1825–32.

96. Biemelt S., Sonnewald U.,Galmbacher P., Willmitzer L., Müller M. Production of human papilloma virus type16virus-like particles in transgenic plants. J Virol 2003,77,9211–20.

97. Kohl T. O., Hitzeroth II,Christensen N. D., Rybicki E. P. Expression of HPV-11L 1proteinin transgenic Arabidopsis thaliana and Nicotiana tabacum. BMC Biotechnol 2007,7,56–69.

98.Soria-GuerraR.E.,Rosales-Mendoza S.,Márquez-Mercado C., López-Revilla R., Castillo-Collazo R., Alpuche-SolísA.G. Transgenic tomatoes express an antigenic polypeptide containing epitopes of the diphtheria, pertussis and tetanus exotoxins, encoded by a syntheticgene. Plant Cell Rep 2007,26, 961–8.

99.Ashraf S.,Singh P. K., Yadav D. K., Shahnawaz M., Mishra S., Sawant S. V., High level expression of surface glycoprotein of rabies virus in tobacco leaves and its immune protective activity in mice.J Biotechnol 2005,119,1-14

100. Arango I. P., Rubio E. L.,Anaya E. R.,FloresT.O.,de la VaraLG,Lim MAG. Expression of the rabies virus nucleo protein in plants at high-levels and evaluation of immune responses in mice. Plant Cell Rep 2008,27,677–85.

101.Yang Z.,Liu Q.,Pan Z.,Yu Z.,Jiao X.Expression of the fusion glycoprotein o new castle disease viru s in transgenic rice and its immune genicity in mice.Vaccine2007,25,591–8.

102.Huang Z., Dry I.,Webster D., Strugne llR.,Wesselingh S. Plant-derived measles virus hemagglutinin protein induces neutralizing antibodies in mice.Vaccine2001,19.

103.Webster D. E., Thomasc M. C., Huang Z., Wesselingh S. L. The development of a plant-based vaccine or measles.Vaccine 2005,23,1859–65.

104.Khandelwal A.,SitaG.L.,Shaila M.S. Oral immunization of cattle with hemagglutin in protein frinder pest virus expressed intransgenic pea nut induces specific immune responses.Vaccine2003,21,3282–9.

105.Portocarrero C. ,Markley K, Koprowski H.,Spitsin S., Golovkin M. Immunogenic properties of plant-derived recombinant smallpox vaccine and idate pB5. Vaccine 2008,26,5535–40.

106. Kim Y S, KangT J, Jang Y S, Yang M S. Expression of neutralizing epitope of porcine epidemic diarrhea virus in potato plants .Plant Cell Tissue Organ Cult2005;82:125–30.

107.Kang T-J,Han S.C.,Yang M.S/, Jang Y.S. Expression of synthetic neutralizing pitope of porcine epidemic diarrhea virus fused with synthetic Bsubunit of Escherichia coli heat-labile entero toxin into bacco plants. Protein ExprPurif 2006a,46,16–22

108.Matsumura T., ItchodaN.,Tsunemitsu H. Production of immunogenic VP6 protein of bovine group A rotavirus intransgenic potato plants.ArchVirol2002;147:1263–70.
109. Wu Y. Z. ,Li J.T., Mou Z.R., Fei L.,Ni B.,Geng M., Oral immunization with rotavirusVP7 expressed in transgenic potatoes induced high titers of mucosal neutralizing IgA. Virology 2003,313 , 337–42.

110.Tomasz Pniewski TheTwenty-Year Story of a Plant-Based Vaccine Against HepatitisB: Stagnation orPromisingProspects?Int.J.Mol.Sci.2013, *14*, 1978-1998.

111.P Lal, V.G. Ramachandran, R. Goyal, R. Sharma. Edible vaccines: Current status and future,2007,2,2,93-102

112.http://www. research and markets.com/research/lwjzhv/analyzing_edible